# THE COPD DIET BIBLE

Beginners' Dietary Plans For Control Of COPD, Recipes And Advice On Lung Health

CRUE GAGE

Copyright © 2024 By Crue Gage

All Rights Reserved.

# Table of Contents

Introductory ................................................. 5

CHAPTER ONE ............................................ 6

  Causes & Risk Factors ............................... 6

  Symptoms & Diagnosis ............................. 8

  Prevention & Treatments Options ............. 10

    Importance Of Nutrition In Managing COPD ................................................... 14

CHAPTER TWO ........................................... 18

  Key Nutrients For Lung Health ................... 18

  Foods To Include And Avoid ...................... 21

  Hydration And Its Importance ................... 24

  7 Days Sample Meal Plans ......................... 26

CHAPTER THREE ......................................... 32

  Managing Weight And COPD ..................... 32

  Addressing Food Allergies And Sensitivities ................................................................ 36

  Importance Of Vitamins And Minerals ...... 39

CHAPTER FOUR ........................................... 43

  Managing Shortness Of Breath During Meals ................................................................ 43

  Tips For Eating Out ................................... 46

- Breakfast Recipes ...................................... 49
- Lunch Recipes .......................................... 54
- Dinner Recipes ......................................... 61
- Snacks And Desserts Recipes ..................... 68

**CHAPTER FIVE** ............................................. 74
- Lifestyle Factors And COPD Management .. 74
- Consulting Healthcare Professionals .......... 79
- Conclusion ............................................... 84

**THE END** .................................................... 86

# Introductory

Chronic Obstructive Pulmonary Disease (COPD) is a progressive lung disease that impedes breathing. It typically encompasses conditions such as emphysema and chronic bronchitis. COPD is frequently the result of prolonged exposure to irritants, including cigarette smoke, air pollution, or chemical vapors, which inflict damage on the lungs and airways.

Coughing, wheezing, chest constriction, and shortness of breath are potential symptoms. The objective of treatment is to alleviate symptoms, delay the progression of the disease, and enhance the quality of life. This is achieved through the use of medications, pulmonary rehabilitation, and, in severe cases, supplemental oxygen or surgery.

# CHAPTER ONE
## Causes & Risk Factors

The causes and risk factors for Chronic Obstructive Pulmonary Disease (COPD) include:

### Causes:

- **Smoking**: The leading cause; both active and passive smoking contribute.
- **Air Pollution**: Long-term exposure to outdoor air pollution or indoor pollutants (e.g., from cooking or heating with biomass fuels).
- **Occupational Exposure**: Jobs that involve exposure to dust, chemicals, or fumes (e.g., construction, mining).

- **Genetic Factors**: A rare genetic condition called alpha-1 antitrypsin deficiency can increase risk.

## Risk Factors:

- **Age**: Risk increases with age, particularly after 40.
- **Gender**: Men are generally more likely to develop COPD, though rates among women are rising.
- **History of Respiratory Infections**: Frequent lung infections in childhood can increase the risk.
- **Family History**: A family history of COPD may increase susceptibility.
- **Asthma**: Having asthma can increase the risk of developing

COPD, especially if it is poorly controlled.

Reducing exposure to these risk factors can help prevent COPD or slow its progression.

## Symptoms & Diagnosis

### Symptoms Of COPD:

- **Chronic Cough**: A persistent cough that produces mucus.
- **Shortness of Breath**: Especially during physical activities; can worsen over time.
- **Wheezing**: A whistling or squeaky sound when breathing.
- **Chest Tightness**: Feeling of pressure or constriction in the chest.

- **Frequent Respiratory Infections**: Increased susceptibility to colds, flu, and other infections.
- **Fatigue**: General tiredness or lack of energy.

**<u>Diagnosis:</u>**

- **Medical History and Symptoms**: A doctor will assess symptoms, smoking history, and exposure to lung irritants.
- **Physical Exam**: Checking for signs like wheezing or prolonged expiration.
- **Spirometry**: A lung function test that measures how much air you can breathe out and how quickly. It's crucial for diagnosing COPD.

- **Chest X-ray or CT Scan**: Imaging tests to check for lung damage and rule out other conditions.
- **Arterial Blood Gas Test**: Measures oxygen and carbon dioxide levels in the blood, assessing lung function.

Diagnosis is typically made based on a combination of these assessments. Early detection and management are vital for improving outcomes.

## Prevention & Treatments Options

### Prevention Of COPD:

- **Avoid Smoking**: The most effective way to prevent COPD; quitting smoking at any age can improve lung health.

- **Reduce Exposure to Pollutants**: Limit exposure to outdoor air pollution and indoor irritants (e.g., secondhand smoke, fumes, dust).
- **Protective Measures**: Use masks and ventilation when working with harmful chemicals or in dusty environments.
- **Healthy Lifestyle**: Maintain a balanced diet and regular exercise to support overall lung and heart health.
- **Vaccinations**: Stay up-to-date with vaccinations (e.g., flu, pneumonia) to prevent respiratory infections.

## **Treatment Options for COPD:**

**Medications:**

- **Bronchodilators:** Help open airways (e.g., albuterol, tiotropium).
- **Inhaled Corticosteroids:** Reduce inflammation (e.g., fluticasone).
- **Combination Inhalers:** Combine bronchodilators and corticosteroids.
- **Phosphodiesterase-4 Inhibitors:** Reduce inflammation and relax airways (e.g., roflumilast).
- **Pulmonary Rehabilitation:** A structured program that includes exercise training, education, and support to improve lung function and overall well-being.

- **Oxygen Therapy**: For those with low blood oxygen levels; supplemental oxygen can help improve quality of life.
- **Surgery**: In severe cases, procedures like lung volume reduction surgery or lung transplantation may be considered.
- **Lifestyle Changes**: Regular exercise, a healthy diet, and stress management can also help manage symptoms and improve quality of life.

Regular follow-up with a healthcare provider is essential for monitoring the disease and adjusting treatment as needed.

## Importance Of Nutrition In Managing COPD

Nutrition plays a vital role in managing Chronic Obstructive Pulmonary Disease (COPD) for several reasons:

### 1. Maintaining Healthy Weight:

- **Underweight**: Low body weight can weaken muscles, including respiratory muscles, making breathing more difficult.

- **Overweight**: Excess weight can increase the workload on the lungs and heart, worsening symptoms.

### 2. Boosting Immune Function:

- A well-balanced diet rich in vitamins and minerals (especially vitamins C and D, zinc, and omega-3 fatty acids) can enhance the immune system, helping to prevent respiratory infections.

### **3. Improving Energy Levels:**

• Proper nutrition provides the energy needed for daily activities and can reduce fatigue, which is common in COPD patients.

### **4. Reducing Inflammation:**

• Antioxidant-rich foods (e.g., fruits, vegetables, nuts) can help combat inflammation in the lungs and improve overall lung function.

### **5. Supporting Muscle Health:**

• Adequate protein intake is crucial for maintaining muscle mass, which is essential for respiratory and physical function.

## 6. Managing Gastroesophageal Reflux Disease (GERD):

Many COPD patients experience GERD, which can worsen respiratory symptoms. A healthy diet can help manage GERD symptoms, improving overall comfort.

## Tips for a COPD-Friendly Diet:

- **Focus on Whole Foods**: Prioritize fruits, vegetables, whole grains, lean proteins, and healthy fats.
- **Stay Hydrated**: Drink plenty of fluids to help thin mucus and make it easier to clear from the lungs.
- **Small, Frequent Meals**: Eating smaller meals can help avoid feelings of fullness that may make breathing more difficult.
- **Limit Salt and Sugar**: Reducing sodium can help manage blood

pressure and limit sugar intake for overall health.

Working with a healthcare provider or nutritionist can help tailor dietary recommendations to individual needs, enhancing overall management of COPD.

# CHAPTER TWO
## Key Nutrients For Lung Health

Several key nutrients are particularly beneficial for lung health, especially for individuals with Chronic Obstructive Pulmonary Disease (COPD):

**1. Antioxidants**:

• **Vitamin C**: Found in citrus fruits, strawberries, and bell peppers; helps protect lung cells from damage.

• **Vitamin E**: Present in nuts, seeds, and green leafy vegetables; supports immune function and reduces inflammation.

**2. Vitamin D**:

• Found in fatty fish, fortified dairy products, and sunlight exposure;

important for immune function and may help reduce respiratory infections.

### **3. Omega-3 Fatty Acids:**

- Found in fatty fish (like salmon), flaxseeds, and walnuts; have anti-inflammatory properties that can benefit lung health.

### **4. Magnesium:**

- Present in nuts, seeds, whole grains, and leafy greens; helps relax the bronchial muscles and can improve lung function.

### **5. Zinc:**

- Found in meat, shellfish, legumes, and seeds; supports immune function and may help reduce the frequency of respiratory infections.

### 6. Fiber:

- Found in fruits, vegetables, whole grains, and legumes; may help reduce inflammation and improve overall health.

### 7. Flavonoids:

- Present in a variety of fruits and vegetables (e.g., berries, onions, and citrus); have antioxidant and anti-inflammatory effects that can benefit lung function.

Incorporating these nutrients through a balanced diet can support lung health and overall well-being. Consulting with a healthcare provider or nutritionist can help create a personalized nutrition plan.

# Foods To Include And Avoid

## Foods to Include for Lung Health:

### Fruits and Vegetables:

- Berries, citrus fruits, spinach, kale, and broccoli for vitamins and antioxidants.

### Whole Grains:

- Brown rice, quinoa, oats, and whole grain bread for fiber and nutrients.

### Lean Proteins:

- Chicken, turkey, fish (especially fatty fish like salmon), legumes, and tofu for muscle support.

### Nuts and Seeds:

- Walnuts, flaxseeds, and chia seeds for omega-3 fatty acids and antioxidants.

**Healthy Fats:**

- Olive oil and avocado for anti-inflammatory benefits.

**Dairy or Dairy Alternatives:**

- Low-fat yogurt and milk (or fortified plant-based alternatives) for calcium and vitamin D.

## Foods To Avoid:

**Processed Foods:**

- High in sugar, salt, and unhealthy fats; can lead to inflammation and weight gain.

**Sugary Foods and Beverages:**

- Sodas, sweets, and pastries; can contribute to obesity and inflammation.

**High-Sodium Foods:**

• Canned soups, processed meats, and salty snacks; can lead to fluid retention and increased blood pressure.

**Fried and Fatty Foods:**

• Fried foods and those high in saturated fats; can worsen inflammation and overall health.

**Dairy Products (for some individuals):**

• While dairy can be beneficial, it can increase mucus production for some people. Monitoring individual responses is key.

Adopting a balanced diet focused on whole, nutrient-rich foods while limiting processed and unhealthy options can support lung health and overall well-being.

## Hydration And Its Importance

Hydration is crucial for overall health and plays a significant role in managing Chronic Obstructive Pulmonary Disease (COPD) for several reasons:

• Proper hydration helps keep mucus in the lungs thin and easier to expel, which can improve breathing and reduce the risk of infections.

• Dehydration can lead to increased mucus viscosity and respiratory difficulties, making it harder to breathe.

• Staying hydrated can help combat fatigue, which is common in COPD patients, enhancing overall energy and quality of life.

- Adequate fluid intake supports kidney function, digestion, and circulation, all of which are important for optimal health.

- Staying hydrated can enhance physical performance, making it easier to engage in pulmonary rehabilitation exercises and daily activities.

**<u>Hydration Tips:</u>**

- Drink water consistently throughout the day.
- Include hydrating foods like fruits (e.g., watermelon, oranges) and vegetables (e.g., cucumbers, lettuce).
- Monitor urine color; pale yellow usually indicates good hydration.
- Limit caffeine and alcohol, as they can contribute to dehydration.

Maintaining proper hydration can significantly impact lung function and overall well-being in individuals with COPD.

## 7 Days Sample Meal Plans

Here's a 7-day sample meal plan that focuses on lung health, incorporating nutrient-rich foods beneficial for individuals with COPD:

### Day 1:

- **Breakfast**: Oatmeal topped with berries and a drizzle of honey.
- **Snack**: Sliced apple with almond butter.
- **Lunch**: Grilled chicken salad with spinach, tomatoes, cucumbers, and olive oil dressing.

- **Snack**: Greek yogurt with chia seeds.
- **Dinner**: Baked salmon with quinoa and steamed broccoli.

## Day 2:

- **Breakfast**: Smoothie with spinach, banana, and flaxseeds.
- **Snack**: Carrot sticks with hummus.
- **Lunch**: Lentil soup with whole grain bread.
- **Snack**: Handful of walnuts.
- **Dinner**: Stir-fried tofu with mixed vegetables and brown rice.

## Day 3:

- **Breakfast**: Scrambled eggs with spinach and whole grain toast.
- **Snack**: Orange slices.

- **Lunch**: Turkey and avocado wrap in a whole grain tortilla with lettuce.
- **Snack**: Celery sticks with peanut butter.
- **Dinner**: Grilled shrimp with asparagus and a side of sweet potatoes.

## Day 4:

- **Breakfast**: Chia pudding made with almond milk topped with sliced bananas.
- **Snack**: Cottage cheese with pineapple.
- **Lunch**: Quinoa salad with chickpeas, bell peppers, and lemon vinaigrette.
- **Snack**: Mixed berries.

- **Dinner**: Baked chicken thighs with roasted Brussels sprouts and brown rice.

## Day 5:

- **Breakfast**: Whole grain pancakes topped with fresh fruit.
- **Snack**: Hard-boiled eggs.
- **Lunch**: Vegetable and bean chili with a side salad.
- **Snack**: Sliced bell peppers with guacamole.
- **Dinner**: Cod fillet with steamed carrots and wild rice.

## Day 6:

- **Breakfast**: Yogurt parfait with granola and mixed berries.
- **Snack**: Sliced cucumber with tzatziki.

- **Lunch**: Spinach and feta stuffed whole grain pita with a side of mixed greens.
- **Snack**: A handful of almonds.
- **Dinner**: Grilled turkey burgers (no bun) with sautéed zucchini and quinoa.

## Day 7:

- **Breakfast**: Smoothie with kale, pineapple, and protein powder.
- **Snack**: Pear slices with cheese.
- **Lunch**: Quinoa and black bean bowl with corn, avocado, and salsa.
- **Snack**: Cherry tomatoes with balsamic vinegar.
- **Dinner**: Roast chicken with mixed roasted vegetables and a side of brown rice.

Remember to drink plenty of water throughout the day, and adjust portions according to individual dietary needs and preferences.

# CHAPTER THREE
## Managing Weight And COPD

Managing weight is crucial for individuals with Chronic Obstructive Pulmonary Disease (COPD) because both underweight and overweight conditions can negatively impact lung function and overall health.

### Importance of Weight Management in COPD:

**Underweight:**

• Can lead to muscle wasting, including respiratory muscles, making breathing more difficult.

• Increases susceptibility to infections and weakens the immune system.

- May result in decreased energy levels and fatigue.

**Overweight:**

- Excess weight can strain the lungs and diaphragm, making it harder to breathe.

- Increases the risk of comorbidities such as heart disease and diabetes.

- Can lead to worsening symptoms and reduced exercise tolerance.

**<u>Strategies for Managing Weight:</u>**

**Balanced Diet:**

- Focus on nutrient-dense foods, including fruits, vegetables, whole grains, lean proteins, and healthy fats.

- Monitor portion sizes and avoid high-calorie, low-nutrient foods.

**Regular Physical Activity**:

- Engage in appropriate exercises, including aerobic activities (like walking) and strength training, as tolerated.

- Consider pulmonary rehabilitation programs for tailored exercise plans.

**Frequent, Small Meals**:

- Eating smaller meals more frequently can help manage appetite and avoid feeling overly full, which can hinder breathing.

**Stay Hydrated**:

- Adequate hydration supports overall health and can help thin mucus, making it easier to breathe.

**Consult Healthcare Professionals:**

• Work with a dietitian or nutritionist for personalized meal plans and strategies.

• Regular check-ups with a healthcare provider can help monitor weight and lung function.

**Mindful Eating:**

• Pay attention to hunger and fullness cues to avoid overeating.

• Keep a food diary to track food intake and identify patterns.

By maintaining a healthy weight, individuals with COPD can improve their lung function, enhance their quality of life, and better manage their symptoms.

# Addressing Food Allergies And Sensitivities

Addressing food allergies and sensitivities is important for individuals with Chronic Obstructive Pulmonary Disease (COPD) as certain foods can trigger respiratory issues or worsen symptoms. Here are key steps to manage food allergies and sensitivities:

• Work with an allergist or dietitian to identify specific food allergies or sensitivities. Track foods consumed and note any adverse reactions, which can help identify patterns.

• Once allergens are identified, eliminate them from your diet. Common allergens include dairy, wheat, nuts, shellfish, and soy. Read labels carefully to avoid hidden allergens in processed foods.

- Find substitutes that provide similar nutrients without triggering allergies (e.g., almond milk for dairy if lactose intolerant). Focus on whole foods like fruits, vegetables, lean proteins, and whole grains that are less likely to cause reactions.

- Be aware of any symptoms that may arise after consuming certain foods, such as respiratory issues, digestive problems, or skin reactions.

- Drinking plenty of water can help reduce mucus production and support overall health.

- Consider working with a nutritionist who can help design a balanced meal plan that avoids allergens while ensuring adequate nutrient intake.

- If you have severe allergies, have an emergency plan in place, including carrying an epinephrine auto-injector if prescribed.

By proactively managing food allergies and sensitivities, individuals with COPD can minimize respiratory issues and improve their overall quality of life.

# Importance Of Vitamins And Minerals

Vitamins and minerals are essential for maintaining overall health and play a particularly important role in managing Chronic Obstructive Pulmonary Disease (COPD). Here's why they matter:

**1. Immune Support:**

- **Vitamin C**: Supports immune function and helps reduce inflammation in the lungs.
- **Vitamin D**: Plays a role in immune regulation and may help lower the risk of respiratory infections.

**2. Anti-Inflammatory Effects:**

- **Vitamin E**: Acts as an antioxidant, protecting lung cells from oxidative stress and inflammation.

- **Omega-3 Fatty Acids**: Found in fish and flaxseeds, they have anti-inflammatory properties that can benefit lung health.

### 3. Muscle Function:

- **Magnesium**: Important for muscle function, including the muscles used for breathing.
- **Protein**: Essential for maintaining muscle mass, which is crucial for respiratory health.

### 4. Bone Health:

- **Calcium and Vitamin D**: Important for bone health, especially as COPD patients may be at higher risk for osteoporosis due to steroid use or reduced physical activity.

## 5. Energy Production:

- **B Vitamins (e.g., B6, B12)**: Help in energy metabolism, which is important for maintaining activity levels and overall vitality.

## 6. Antioxidant Protection:

- **Zinc**: Supports immune function and has antioxidant properties that can help protect lung tissue.

## 7. Overall Health Maintenance:

- A balanced intake of vitamins and minerals contributes to better management of comorbidities often associated with COPD, such as heart disease and diabetes.

Incorporating a variety of nutrient-rich foods, such as fruits, vegetables, whole grains, lean proteins, and healthy fats, ensures adequate intake of essential vitamins and minerals. If needed, supplements should be discussed with a healthcare provider to avoid excessive intake and potential interactions with medications.

# CHAPTER FOUR
## Managing Shortness Of Breath During Meals

Managing shortness of breath during meals can be challenging for individuals with Chronic Obstructive Pulmonary Disease (COPD). Here are some strategies to help:

• Consuming smaller portions more frequently can prevent feelings of fullness that may hinder breathing.

• Focus on high-nutrient foods that provide essential vitamins and minerals without requiring large portions (e.g., smoothies, soups).

• Fatty or heavy meals can make breathing more difficult, so opt for lighter options that are easier to digest.

- Maintain an upright posture during meals to allow for better lung expansion and easier breathing.

- Eat slowly and chew thoroughly to reduce the need for quick, deep breaths that can exacerbate shortness of breath.

- Minimize conversation during meals to focus on breathing and reduce the likelihood of becoming winded.

- Drink fluids between meals rather than during meals to avoid feeling too full and to help thin mucus.

- Practice pursed-lip breathing or diaphragmatic breathing before and during meals to help manage shortness of breath.

- Schedule meals during times of day when you feel most energetic and least breathless.

- Prepare meals in advance or use meal delivery services to save energy on cooking.

By implementing these strategies, individuals with COPD can improve their eating experience and manage shortness of breath more effectively during meals.

## Tips For Eating Out

Eating out can be enjoyable and manageable for individuals with COPD. Here are some tips to make the experience easier and healthier:

• Look for places that offer a variety of healthy options, including salads, grilled items, and whole grains.

• Check the menu online before going to the restaurant to plan your meal and avoid feeling rushed.

• Don't hesitate to ask questions about the menu items or request modifications (e.g., less salt, no heavy sauces).

• Consider sharing an entrée or ordering an appetizer as a main dish to avoid overeating.

- Choose grilled, baked, or steamed options over fried foods to make meals easier on your digestive system.

- Drink water throughout the meal to stay hydrated without feeling overly full.

- Take your time while eating, savoring each bite to aid digestion and help manage shortness of breath.

- Choose quieter times to dine out when restaurants are less crowded, reducing stress and allowing for easier breathing.

- Be mindful of how certain foods affect your breathing; avoid known triggers or heavy foods that can exacerbate symptoms.

- Eating with a friend or family member can provide support and make the experience more enjoyable.

By following these tips, individuals with COPD can have a more pleasant and healthier dining experience while managing their symptoms effectively.

# Breakfast Recipes

Here are some healthy breakfast recipes that are great for individuals with COPD:

## **1. Spinach and Feta Omelette:**

**Ingredients**:

- 2 eggs
- Handful of fresh spinach
- 1/4 cup feta cheese, crumbled
- Salt and pepper to taste
- Olive oil or cooking spray

**Instructions**:

- Heat a non-stick skillet over medium heat and add a little olive oil.
- Whisk the eggs in a bowl, adding salt and pepper.

- Pour the eggs into the skillet and cook for a minute.
- Add spinach and feta on one half of the omelet.
- Fold the omelet and cook until fully set. Serve warm.

## 2. Berry Oatmeal:

**Ingredients**:

- 1/2 cup rolled oats
- 1 cup water or milk (dairy or plant-based)
- 1/2 cup mixed berries (fresh or frozen)
- 1 tablespoon honey or maple syrup (optional)
- A sprinkle of cinnamon

**Instructions:**

- In a saucepan, bring water or milk to a boil.
- Add oats and reduce heat to a simmer. Cook for about 5 minutes, stirring occasionally.
- Stir in berries and cook for an additional 2 minutes.
- Sweeten with honey or syrup, if desired, and sprinkle with cinnamon before serving.

### 3. Avocado Toast:

**Ingredients:**

- 1 slice whole grain bread
- 1/2 ripe avocado
- Salt and pepper to taste
- Optional toppings: sliced tomatoes, radishes, or a poached egg

**Instructions**:

- Toast the bread to your liking.
- Mash the avocado in a bowl and season with salt and pepper.
- Spread the avocado mixture on the toast and add any optional toppings.
- Serve immediately.

### 4. Chia Seed Pudding:

**Ingredients**:

- 1/4 cup chia seeds
- 1 cup almond milk (or any milk of choice)
- 1 tablespoon honey or maple syrup (optional)
- Fresh fruit for topping (e.g., bananas, berries)

**Instructions**:

- In a bowl or jar, mix chia seeds, almond milk, and sweetener.
- Stir well and let sit for about 10 minutes, then stir again to prevent clumping.
- Refrigerate for at least 2 hours or overnight.
- Serve chilled, topped with fresh fruit.

**<u>5. Banana Smoothie:</u>**

**Ingredients**:

- 1 ripe banana
- 1 cup spinach (optional)
- 1/2 cup Greek yogurt
- 1 cup almond milk (or any milk of choice)
- 1 tablespoon nut butter (optional)

**Instructions**:

- Combine all ingredients in a blender.
- Blend until smooth, adding more milk for desired consistency.
- Pour into a glass and enjoy immediately.

These recipes are nutritious, easy to prepare, and can help provide energy while supporting overall health.

## Lunch Recipes

Here are some healthy lunch recipes that are great for individuals with COPD:

### 1. Quinoa Salad with Chickpeas:

**Ingredients**:

- 1 cup cooked quinoa

- 1 can chickpeas, rinsed and drained
- 1/2 cup cherry tomatoes, halved
- 1/4 cup cucumber, diced
- 1/4 cup red onion, finely chopped
- 2 tablespoons olive oil
- Juice of 1 lemon
- Salt and pepper to taste
- Fresh parsley or cilantro for garnish

**Instructions**:

- In a large bowl, combine quinoa, chickpeas, tomatoes, cucumber, and onion.
- In a small bowl, whisk together olive oil, lemon juice, salt, and pepper.
- Pour the dressing over the salad and toss gently.

- Garnish with parsley or cilantro before serving.

## **2. Turkey and Avocado Wrap:**

**Ingredients**:

- 1 whole grain tortilla
- 4-6 slices of turkey breast
- 1/2 avocado, sliced
- Handful of spinach or lettuce
- Sliced tomato
- Mustard or hummus (optional)

**Instructions**:

- Lay the tortilla flat and spread mustard or hummus if using.
- Layer turkey, avocado, spinach, and tomato on top.
- Roll up the tortilla tightly and slice in half. Serve immediately.

### **3. Lentil Soup**

**Ingredients**:

- 1 cup dried lentils, rinsed
- 1 onion, chopped
- 2 carrots, diced
- 2 celery stalks, diced
- 3 cloves garlic, minced
- 4 cups vegetable or chicken broth
- 1 teaspoon cumin
- Salt and pepper to taste
- Olive oil for sautéing

**Instructions**:

- In a pot, heat olive oil over medium heat. Add onion, carrots, celery, and garlic; sauté until softened.
- Stir in lentils, broth, cumin, salt, and pepper. Bring to a boil.

- Reduce heat and simmer for 25-30 minutes until lentils are tender.
- Serve warm with crusty whole grain bread.

## 4. Baked Sweet Potato with Black Beans:

**Ingredients**:

- 1 medium sweet potato
- 1 can black beans, rinsed and drained
- 1/2 avocado, diced
- Salsa (optional)
- Cilantro for garnish (optional)

**Instructions**:

- Preheat the oven to 400°F (200°C). Pierce the sweet potato with a fork and bake for 45-60 minutes until tender.

- Once cooked, let cool slightly, then slice open.
- Top with black beans, avocado, and salsa if desired. Garnish with cilantro before serving.

## **5. Mediterranean Tuna Salad:**

**Ingredients**:

- 1 can tuna (in water), drained
- 1/4 cup cherry tomatoes, halved
- 1/4 cup cucumber, diced
- 1/4 cup red onion, chopped
- 2 tablespoons olives, sliced
- 2 tablespoons olive oil
- Juice of 1 lemon
- Salt and pepper to taste
- Fresh basil or parsley for garnish

**Instructions**:

- In a bowl, combine tuna, tomatoes, cucumber, onion, and olives.
- In a small bowl, whisk together olive oil, lemon juice, salt, and pepper.
- Pour the dressing over the salad and toss gently.
- Garnish with fresh herbs before serving.

These lunch recipes are nutritious, easy to prepare, and packed with flavor, making them great choices for maintaining energy and supporting lung health.

# Dinner Recipes

Here are some healthy dinner recipes that are great for individuals with COPD:

## 1. Grilled Lemon Herb Chicken:

**Ingredients**:

- 2 boneless, skinless chicken breasts
- Juice of 1 lemon
- 2 tablespoons olive oil
- 2 cloves garlic, minced
- 1 teaspoon dried oregano
- Salt and pepper to taste

**Instructions**:

- In a bowl, mix lemon juice, olive oil, garlic, oregano, salt, and pepper.

- Marinate the chicken in the mixture for at least 30 minutes.
- Preheat the grill or a grill pan over medium heat. Grill chicken for 6-7 minutes on each side or until cooked through.
- Serve with a side of steamed vegetables.

## 2. Baked Salmon with Asparagus

**Ingredients**:

- 2 salmon fillets
- 1 bunch asparagus, trimmed
- 2 tablespoons olive oil
- Juice of 1 lemon
- Salt and pepper to taste
- Fresh dill or parsley for garnish

**Instructions**:

- Preheat the oven to 400°F (200°C).

- On a baking sheet, arrange salmon and asparagus. Drizzle with olive oil, lemon juice, salt, and pepper.
- Bake for 12-15 minutes until salmon is cooked through and asparagus is tender.
- Garnish with fresh herbs before serving.

### 3. Vegetable Stir-Fry with Tofu

**Ingredients**:

- 1 block firm tofu, pressed and cubed
- 2 cups mixed vegetables (bell peppers, broccoli, carrots, snap peas)
- 2 tablespoons soy sauce (low sodium)
- 1 tablespoon sesame oil
- 2 cloves garlic, minced

- Cooked brown rice or quinoa (for serving)

**Instructions:**

- Heat sesame oil in a large skillet or wok over medium heat. Add tofu and sauté until golden brown.
- Add garlic and mixed vegetables; stir-fry for about 5-7 minutes until vegetables are tender-crisp.
- Stir in soy sauce and cook for an additional 2 minutes.
- Serve over brown rice or quinoa.

## **4. Quinoa Stuffed Bell Peppers**

**Ingredients:**

- 2 large bell peppers, halved and seeds removed
- 1 cup cooked quinoa

- 1 can black beans, rinsed and drained
- 1/2 cup corn (fresh or frozen)
- 1 teaspoon cumin
- Salt and pepper to taste
- 1/2 cup salsa

**Instructions**:

- Preheat the oven to 375°F (190°C).
- In a bowl, combine cooked quinoa, black beans, corn, cumin, salt, pepper, and salsa.
- Stuff the mixture into the halved bell peppers.
- Place stuffed peppers in a baking dish and cover with foil. Bake for 30-35 minutes until peppers are tender.

## **5. Shrimp Tacos with Cabbage Slaw**

**Ingredients**:

- 1 pound shrimp, peeled and deveined
- 1 tablespoon olive oil
- 1 teaspoon chili powder
- Salt and pepper to taste
- Corn or whole grain tortillas
- 2 cups shredded cabbage
- 1 tablespoon lime juice
- Fresh cilantro for garnish

**Instructions**:

- In a skillet, heat olive oil over medium heat. Season shrimp with chili powder, salt, and pepper. Cook shrimp for 2-3 minutes on each side until pink and cooked through.

- In a bowl, toss shredded cabbage with lime juice and a pinch of salt.
- Warm tortillas and fill with shrimp and cabbage slaw. Garnish with fresh cilantro before serving.

These dinner recipes are nutritious, easy to prepare, and delicious, making them perfect for supporting lung health and overall well-being.

# Snacks And Desserts Recipes

Here are some healthy snack and dessert recipes that are great for individuals with COPD:

## **Healthy Snacks**

### **1. Hummus with Veggie Sticks:**

**Ingredients**:

- 1 can chickpeas, rinsed and drained
- 2 tablespoons tahini
- 2 tablespoons olive oil
- Juice of 1 lemon
- 1 clove garlic, minced
- Salt and pepper to taste
- Assorted veggie sticks (carrots, celery, cucumber, bell peppers)

**Instructions**:

- In a blender, combine chickpeas, tahini, olive oil, lemon juice, garlic, salt, and pepper. Blend until smooth, adding water for desired consistency.
- Serve with fresh veggie sticks.

## 2. Greek Yogurt Parfait

**Ingredients**:

- 1 cup Greek yogurt (plain or flavored)
- 1/2 cup mixed berries (fresh or frozen)
- 1/4 cup granola or nuts (optional)
- Honey or maple syrup (optional)

**Instructions:**

- In a glass or bowl, layer Greek yogurt, berries, and granola or nuts.
- Drizzle with honey or syrup if desired. Serve immediately.

## Healthy Desserts

### 3. Chia Seed Pudding:

**Ingredients:**

- 1/4 cup chia seeds
- 1 cup almond milk (or any milk of choice)
- 1 tablespoon honey or maple syrup (optional)
- Fresh fruit for topping (e.g., bananas, berries)

**Instructions**:

- In a bowl or jar, mix chia seeds, almond milk, and sweetener.
- Stir well and let sit for about 10 minutes, then stir again to prevent clumping.
- Refrigerate for at least 2 hours or overnight.
- Serve chilled, topped with fresh fruit.

## 4. Baked Apples with Cinnamon

**Ingredients**:

- 2 apples, cored and halved
- 2 tablespoons oats
- 2 tablespoons chopped nuts (e.g., walnuts, pecans)
- 1 teaspoon cinnamon
- Honey (optional)

**Instructions**:

- Preheat the oven to 350°F (175°C).
- In a bowl, mix oats, nuts, and cinnamon. Fill the center of each apple half with the mixture.
- Place apples in a baking dish and drizzle with honey if desired.
- Bake for 20-25 minutes until tender. Serve warm.

## **5. Frozen Banana Bites:**

**Ingredients**:

- 2 ripe bananas
- 1/2 cup dark chocolate chips (or yogurt coating)
- Chopped nuts or coconut (optional)

**Instructions**:

- Slice bananas into bite-sized pieces.
- Melt chocolate chips in a microwave or double boiler until smooth.
- Dip each banana slice in chocolate and place on a parchment-lined baking sheet. Sprinkle with nuts or coconut if desired.
- Freeze for at least 1 hour until set. Enjoy frozen!

These snack and dessert recipes are nutritious, delicious, and easy to prepare, making them perfect for satisfying cravings while supporting overall health.

# CHAPTER FIVE
## Lifestyle Factors And COPD Management

Managing Chronic Obstructive Pulmonary Disease (COPD) involves addressing various lifestyle factors that can significantly impact symptoms and overall health. Here are key lifestyle considerations:

**1. Smoking Cessation:**

• **Importance**: Smoking is the leading cause of COPD. Quitting can slow disease progression and improve lung function.

• **Tips**: Seek support through cessation programs, counseling, or medications like nicotine replacement therapy.

## 2. Healthy Diet:

- **Importance**: Proper nutrition supports lung health and overall well-being.

- **Tips**: Focus on a balanced diet rich in fruits, vegetables, whole grains, lean proteins, and healthy fats. Stay hydrated.

## 3. Regular Exercise:

- **Importance**: Physical activity improves lung function, reduces breathlessness, and enhances overall stamina.

- **Tips**: Engage in aerobic exercises (walking, cycling) and strength training. Consider pulmonary rehabilitation programs for tailored exercise plans.

### 4. Weight Management:

- **Importance**: Maintaining a healthy weight can help reduce strain on the lungs and improve breathing.

- **Tips**: Monitor diet and engage in regular physical activity to achieve and maintain a healthy weight.

### 5. Stress Management:

- **Importance**: Stress can exacerbate COPD symptoms.

- **Tips**: Practice relaxation techniques such as deep breathing, meditation, or yoga. Engage in hobbies that promote relaxation and joy.

### 6. Avoiding Respiratory Irritants:

- **Importance**: Exposure to pollutants, allergens, and irritants can worsen COPD symptoms.

- **Tips**: Avoid secondhand smoke, limit exposure to dust and fumes, and maintain good indoor air quality.

### 7. Regular Medical Check-ups:

- **Importance**: Routine check-ups help monitor lung function and adjust treatment plans as necessary.

- **Tips**: Stay up-to-date with vaccinations (like flu and pneumonia vaccines) and discuss any new symptoms with your healthcare provider.

**8. Oxygen Therapy**:

- **Importance**: For those with low blood oxygen levels, supplemental oxygen can improve quality of life and exercise tolerance.

- **Tips**: Follow medical advice on the use of oxygen therapy as prescribed.

**9. Stay Informed**:

- **Importance**: Understanding COPD and its management empowers patients to make informed decisions about their health.

- **Tips**: Educate yourself about the condition, treatment options, and management strategies.

By focusing on these lifestyle factors, individuals with COPD can effectively manage their symptoms, improve their quality of life, and slow disease progression. Collaborating with healthcare professionals is essential for personalized care and support.

## Consulting Healthcare Professionals

Consulting healthcare professionals is a crucial aspect of managing Chronic Obstructive Pulmonary Disease (COPD). Here are some key points to consider regarding consultations:

**1. Primary Care Physician**:

• **Role**: Regular check-ups and monitoring of overall health.

• **Tips**: Discuss symptoms, treatment options, and any new concerns during visits.

**2. Pulmonologist**:

- **Role**: Specialist in respiratory conditions, providing expert care for COPD.

- **Tips**: Seek referrals for a pulmonologist for advanced treatment, pulmonary function tests, and specialized therapies.

**3. Registered Dietitian**:

- **Role**: Nutrition expert who can provide tailored dietary advice.

- **Tips**: Consult for meal planning, managing weight, and ensuring adequate nutrition to support lung health.

## 4. Respiratory Therapist:

• **Role**: Assists with breathing techniques, pulmonary rehabilitation, and the use of inhalers and oxygen therapy.

• **Tips**: Work with a respiratory therapist for guidance on managing breathlessness and improving lung function.

## 5. Physical Therapist:

• **Role**: Helps develop safe exercise routines to improve endurance and strength.

• **Tips**: Consider physical therapy as part of pulmonary rehabilitation to enhance overall fitness.

### 6. Allergist/Immunologist:

- **Role**: Evaluates and treats allergies that may exacerbate COPD symptoms.

- **Tips**: Discuss any known allergies or asthma-like symptoms that could affect lung health.

### 7. Mental Health Professional:

- **Role**: Supports mental and emotional well-being, addressing anxiety or depression related to COPD.

- **Tips**: Seek counseling or therapy if experiencing stress, anxiety, or depression due to chronic illness.

A comprehensive and effective management plan for COPD can be achieved through regular consultations with these healthcare professionals.

Achieving optimal results necessitates candid discussions regarding symptoms, concerns, and treatment preferences. Always feel confident in your ability to ask inquiries and seek assistance when necessary.

# Conclusion

The management of Chronic Obstructive Pulmonary Disease (COPD) necessitates a multifaceted approach that encompasses effective symptom management, regular consultations with healthcare professionals, dietary considerations, and lifestyle modifications.

Quitting smoking, adhering to a nutritious diet, participating in consistent physical activity, managing tension, and remaining informed about the condition are among the most effective strategies.

It is essential to collaborate with healthcare providers, including pulmonologists, dietitians, and respiratory therapists, in order to ensure that patients receive personalized treatment and receive optimal management. Individuals with COPD can enhance their quality of life, alleviate symptoms, and delay the progression of the disease by actively engaging in their health care and making informed decisions.

It is imperative that individuals are motivated to seek assistance and resources when necessary, and empowerment through knowledge and support is essential.

**THE END**

www.ingramcontent.com/pod-product-compliance
Lightning Source LLC
Chambersburg PA
CBHW070204230526
45471CB00002B/820